BEI GRIN MACHT SICH IHR WISSEN BEZAHLT

Bibliografische Information der Deutschen Nationalbibliothek:

Die Deutsche Bibliothek verzeichnet diese Publikation in der Deutschen National-
bibliografie; detaillierte bibliografische Daten sind im Internet über http://dnb.d-
nb.de/ abrufbar.

Impressum:

Copyright © 2015 GRIN Verlag, Open Publishing GmbH
Druck und Bindung: Books on Demand GmbH, Norderstedt Germany
ISBN: 978-3-668-18271-4

Dieses Buch bei GRIN:

http://www.grin.com/de/e-book/317908/the-execution-of-a-green-roof-design-
analysis-assessment-concept-and

Marvin Namanda

The Execution of a Green Roof Design. Analysis, Assessment, Concept and Design

GRIN Verlag

Green Roof Design

Part 1: Client Brief and Site Analysis

Client Brief

- The client requires a specially engineered rooftop that supports plant life, measuring 37m by 44m.
- They requested for a rooftop with a thin layer of low-growing, herbaceous plants that require low maintenance, thus an extensive green roof. The roof had to be a multi-layered build up consisting of a protection fleece layer and a re-cycled filter fleece under a loose blend of organic and mineral substrate under a planted sedum vegetation.
- Given the Australian climate, the plants should be densely spaced and able to live in meager soil and also survive exposure to cold, heat, and wind.

Anticipated Outcomes for the green roof

1. Storm water management
2. Reduction in the "Heat Island Effect"
3. Reduced cases of air and noise pollution
4. Extended life of the rooftops
5. Provision of a heating and cooling mechanism for the building
6. An attractive rooftop as an alternative to the traditional rooftop

Feasibility Analysis of the Outcomes

The increasing global concern for the environment has led to the development of environmentally friendly constructions. The adoption of the green roof technology is perhaps a solution to the environmental impacts of a building. The Collingwood-based Robin's Landscape Designers Company is considering installing a green roof on part of the new Robin Center in a bid to enhance the company's image as an environmental conscious. The company has thus enlisted the assistance of our consulting team to investigate the feasibility and the eco-social-economic impacts of the project.

Storm Water Management

According to the United Nations (2002), storm water management is a term used to describe the control of discharge and runoff resulting from rainfall on a particular area. Urban areas like Collingwood make use of storm water management due to their increased impermeability to rainfall infiltration as a function of roads, roof, and parking lots among others. The rationale of storm water management is to attenuate the increased runoff to conditions and levels that the receiving watercourse is capable of handling. Green roof technology is an effective Storm Water Management control method.

Storm water management is subdivided into two categories: water quality control in terms temperature, bacteria, dissolved minerals and suspended solids, and erosion control. Green roofs offer a unique storm water management opportunity especially in the urban areas where land is scarce for the traditional storm water management facilities. Reduction in the volume of storm water runoff from green roofs will reduce flow duration and velocity; hence reduce the erosion on the immediate and surrounding environment of Robins Center. Consequently, the reduction in erosion results in improved water quality as the suspended soil material within the flow will be reduced in quantity. This also improves the town's drainage systems stability and reduces the frequency of their maintenance and rehabilitation. According to Marinova, Annandale, and Phillimore (2006), a large percent of Phosphorous is carried though air and this amount co-relates to the suspended solids. Studies have revealed that green roofs absorb these minerals to levels recommended by the Ministry of Environment with respect to storm management (Graham, 2002).

Air and Noise Pollution Reduction

According to the United Nations Environment Program (2002), Melbourne was ranked among the worst communities by the amount of tropospheric ozone and the amount of airborne particulate matter. It is no doubt that the components of air pollution pose serious health risks to human life. For example, fine particulates from aerosols, smoke, dust, and fumes penetrate the respiratory system calling for medical attention. The study by Werthmann and Michael (2007) suggests that green roofs can remove upto 0.2kg of particulate matter per year across a square meter area of the roof. Research information also reveals that about 31% of the phosphorous source is airborne, therefore, green roofs have the potential to reduce phosphorous. Despite the fact that it is difficult to quantify the air improvement quality by green roof technology, the benefits suggest a better air quality evidenced by the reduced frequency of people seeking medical assistance in the respective areas.

The various layers used in developing a green roof significantly contribute to noise reduction. The *Captains Flat (Abatement of Pollution) Agreement Act 1975*, states that the soil used to construct green roofs absorbs common outdoor noise and the traffic noise too. This benefit is applicable to Robin's Center that is close to a highway.

Green roofs regulate the temperature of the building and, therefore, energy efficiency is achieved. Green roofs improve energy efficiency in a number of ways. First, the roof medium ensures a thermal break from the structural portion of the roof. More so, this thermal break is enhanced by the addition of the insulation layer as it is a requirement for constructing an extensive green roof. The second way through which green roofs improve energy efficiency is through the heat island effect; the vegetative layer gains less heat than the gravel surfaces.

Reducing the Urban Heat Island

On warm summer days in Australia, the air temperature in the suburbs of Melbourne can be several degrees higher than in the less urbanized areas; this phenomenon is known as the Urban Heat Island (Werthmann & Michael, 2007). The increased temperature is associated to the increased amount of roofing and paving, and a reduced size of vegetative cover that normally provide heat attenuation. The high temperatures result to an increase in the energy costs due to the demand for cooling equipment. Consequently, the rise in demand for hydropower increases air pollution through the burning of fuels. There is a scientific consensus that increased vegetation decreases air temperature. Green roofs have an overall impact of a vegetated urban surface, thus a decrease in the Urban Heat Island effect.

Extension of roof life span

The Australian climate cycles require that the infrastructure is designed and constructed with special consideration to temperature changes. Temperature variation affects exterior structures, buildings, sewers, and roadways among others (Graham, 2002). Foundations need frost protection because of latent differential movement of soils during freezing temperatures, therefore, structures should be designed to avoid temperature impacts and allow for expansion and contraction. In the Australian climate, materials expand and contract by up to 10% of their original sizes depending on the type of the material (Graham, 2002). Building roofs has dissimilar contraction and expansion coefficients to the adjoining rate of change. When traditional roofing materials are exposed to temperature changes over time, their elasticity reduces and consequently the roof gets torn. On the contrary, green roofs offer protection from solar radiation and other weather conditions, thus significantly reduce surface degradation and the expansion/contraction cycle. According to the study by Werthmann and Michael (2007), green roofs have the capability to extend the life expectancy of the traditional roofs by a factor of three.

Attractive Roof top

Undoubtedly, green roofs are credited with the ability to improve the beauty of buildings and, as a result, appeal to the senses of people.

Assessment of the Environmental, Social, and Economic Impacts of the Green Roof Technology

Introduction

This section identifies the positive and negative impacts associated with the green roof project. The impacts are identified in two phases of the project: Construction Phase and Operation Phase.

Construction Phase

Negative Impacts

Ecological imbalances

Without appropriate management measures, the project has the potential to create serious ecological imbalances at the project site. The introduction of Sedum grass can affect the microclimate as it has a high water uptake rate. Sedum grass could also oust the indigenous species or introduce disease agents that may affect the plants, animals, and human beings in Collingwood suburb. Pesticides and fertilizers are used to contain the imbalances. These may percolate through the soil and be carried away in the drainage system, thus polluting the ground and surface waters. More so, the nutrients in the fertilizers may result to eutrophication of the surface water bodies and enhance the growth of aquatic weeds. Residues of pesticides are harmful to human and animal health.

Solid waste generation

Large amounts of solid waste will be generated during construction of the green roof. These will include rejected materials, metal cuttings, surplus materials, empty cartons, paper bags, broken plastics, among others. If not well-managed, solid wastes have the potential of causing disease outbreaks as they provide suitable breeding conditions for disease vectors such as mosquitoes and bacteria. The major vulnerable groups are children.

The workers on site will generate fecal matter during the day-to-day operations. This generated waste needs proper handling to prevent diseases such as cholera and diarrhea outbreak on site. Unless addressed adequately, it can prove to be a serious environmental hazard.

Noise and Air Pollution

Particulate natter pollution is likely to occur during the construction of the various roof layers and transportation of waste. There is a possibility of these particulates to affect the site workers and the surrounding set-ups.

Positive Impacts

Employment Opportunities

During the construction phase, availability of employment opportunities is perhaps the main positive impact. These employment opportunities benefit the social and economic spheres of the society. In the economic sphere, abundant unskilled labor will be employed during the construction phase hence economic growth. Several workers including carpenters, masons, joiners, and plumbers are expected to work at the site from the start to the end of the project.

Improving the growth of the economy

The use of locally available materials during the construction phase including cement, concrete, structural steel, and plywood will contribute towards the growth of the economy by improving the gross domestic product.

Operation Phase Impacts

Negative Impacts

Increased water use

Irrigation of the roof could involve significant quantities of water, thus affecting the volumes of commercial water.

Positive Impacts

Improved quality of air

Enhanced Storm Water Management

Improved aesthetic value of the building

Part 2: Concept Design

Introduction

The initial green roof design was a combination of the standard concepts. The design followed the results received from the client's brief and public survey as to what was perceived as important in the site:

1. Water quality
2. Air quality
3. Energy consumption and recreation

The direction given to the consultants included consideration of technical and habitat requirements.

Technical and Habitat Considerations

The various types of greenhouses and suitable species were considered. The potential roof types were based on the natural ecosystems and in conjunction with the Australian practices and guidelines for vegetative species with respect to extensive green roofs:

Habitat	Botanical Name	Common Name
Succulent Ground Creepers	*Tetragonia implexicoma*	Bower Spinach
	Carpobrotus rossii	Karkalla
	Enchylaena tomentosa	Ruby Saltbush
Grasses and Tussocks	*Poa poiformis*	Blue Tussock-Grass
	Dictichlis distichophyiia	Australian Salt-Grass
	Themeda triandra	Kangaroo Grass
	Spinix sericeus	Hairy Spinifex
Climbers	*Nitraria billardierei*	Nitre Bush
	Zygophyllum billardierei	Coast Twin-leaf
Ground Creepers	*Myoporum parvifolium*	Creeping Boobialla
	Scaevoia albida	Small Fan-Flower
	Scaevoia aemula	Fairy Fan-Flower
	Scaevoia calendualacea	Dune Fan-flower

Part 3: Final Design

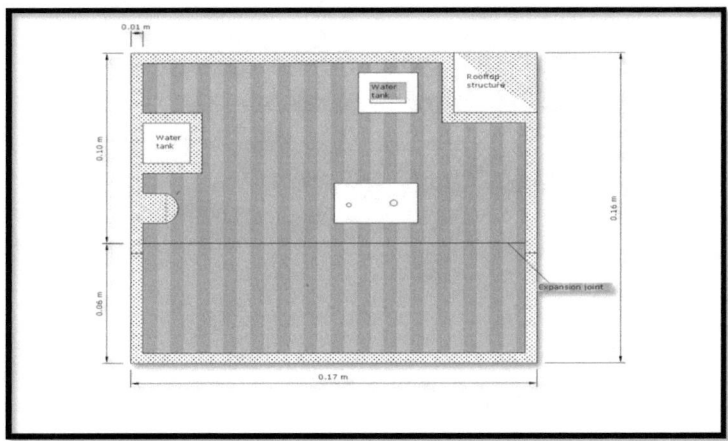

Figure 1 Plan View Scale 1:76

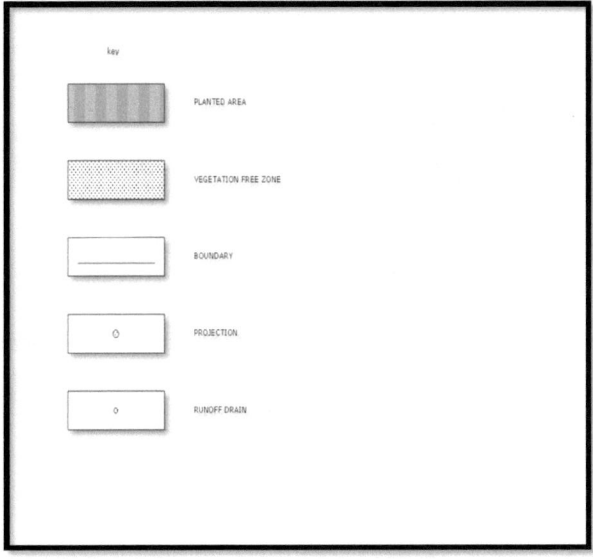

Figure 2 Plan View Key

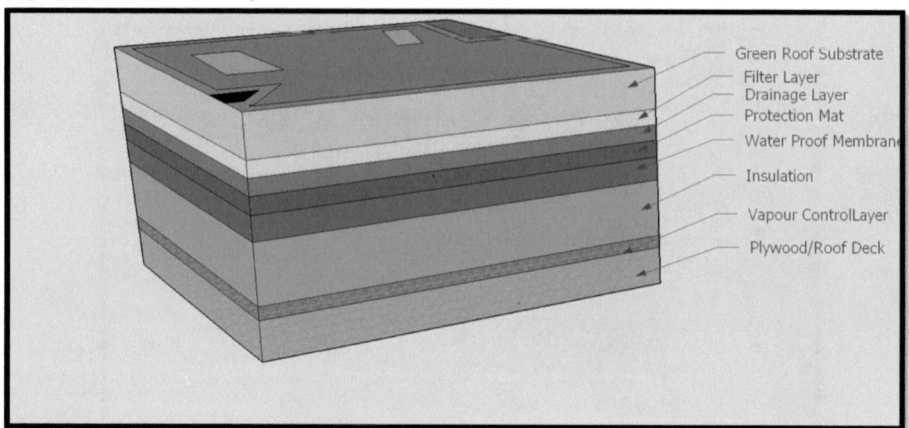

Green Roof Substrate
Filter Layer
Drainage Layer
Protection Mat
Water Proof Membran
Insulation
Vapour ControlLayer
Plywood/Roof Deck

Figure 3 Horizontal View of the extensive green roof's material layers

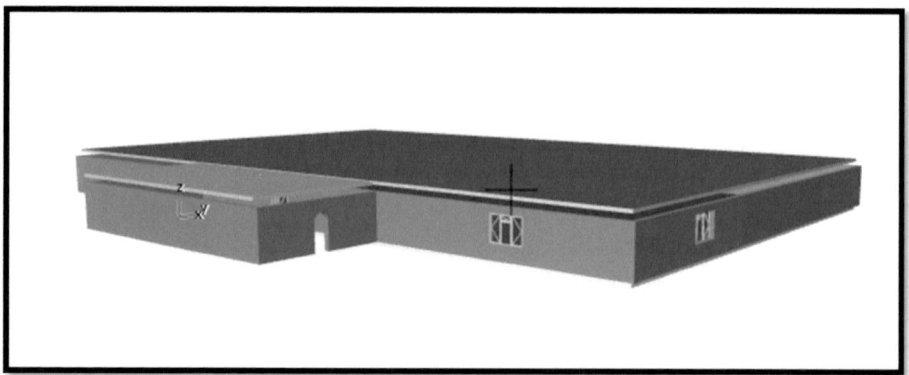

Figure 4 Final Model of the green roof on the building

Planting Schedule

Plant sedum seeds in early spring. The soild should be well-drained and in full sun. The plants should be spaced 6 inches between and 2 feet apart.

Plant description and requirements

Botanical name	Plant Type	USDA Hardiness Zone	Sun Exposure	Soil Type	Flower Color	Bloom Time
Sedum	Flower	3,4,5,6,7	Full Sun	Sandy, Loamy	Red, Yellow, White	Spring, Summer

Maintenance Schedule for Robin's Center Green Roof Systems

Green roofs are living structures, thus need attention throughout their lifespan. Sedum, the chosen vegetation for the green roof, requires less maintenance. Nevertheless, regular attention to soil moisture and feed levels and the removal of weeds is fundamental to maintain optimum performance.

Spring Schedule

- Weeding by hand-pulling and use of herbicides
- Apply slow release granular fertilizer at a rate of 45gsm
- Remove debris clogging the drainage outlets
- Inspect the vegetation growing area and check for poor growth areas
- Check the irrigation system and replace the necessary parts.

Autumn Schedule

- Remove weed only by hand-pulling (Cantor, 2008)
- Clear the drainage outlets
- Inspect the vegetation and note any poor growth areas
- Over-seed the poor growth areas
- Check the irrigation system and drain down

References

Cantor, S. L. (2008). *Green roofs in sustainable landscape design*. New York: W.W. Norton &

    Co.

Graham, P. (2002). *Building ecology: First principles for a sustainable built environment*. Oxford: Blackwell Science.

Marinova, D., Annandale, D., & Phillimore, J. (2006). *The international handbook on environmental technology management*. Cheltenham, UK: Edward Elgar.

United Nations Environment Programme. (2002). *Global environment outlook*. London: Earthscan.

Werthmann, C., & Michael, V. (2007). *Green roof: A case study*. New York: Princeton Architectural Press.

BEI GRIN MACHT SICH IHR WISSEN BEZAHLT

- Wir veröffentlichen Ihre Hausarbeit,
  Bachelor- und Masterarbeit

- Ihr eigenes eBook und Buch -
  weltweit in allen wichtigen Shops

- Verdienen Sie an jedem Verkauf

**Jetzt bei www.GRIN.com hochladen
und kostenlos publizieren**